ANTIQUES AND THEIR VALUES

GLASS

Compiled by
TONY CURTIS

D1080376

Printed by Apollo Press, Unit 5, Dominion Way, Worthing, Sussex.

INTRODUCTION

Congratulations! You now have in your hands an extremely valuable book. It is one of a series specially devised to aid the busy professional dealer in his everyday trading. It will also prove to be of great value to all collectors and those with goods to sell, for it is crammed with illustrations, brief descriptions and valuations of hundreds of antiques.

Every effort has been made to ensure that each specialised volume contains the widest possible variety of goods in its particular category though the greatest emphasis is placed on the middle bracket of trade goods rather than on those once - in - a - lifetime museum pieces whose values are of academic rather than practical interest to the vast majority of dealers and collectors.

This policy has been followed as a direct consequence of requests from dealers who sensibly realise that, no matter how comprehensive their knowledge, there is always a need for reliable, up-to-date reference works for identification and valuation purposes.

When using your Antiques and their Values to assess the worth of goods, please bear in mind that it would be impossible to place upon any item a precise value which would hold good under all circumstances. No antique has an exactly calculable value; its price is always the result of a compromise reached between buyer and seller, and questions of condition, local demand and the business acumen of the parties involved in a sale are all factors which affect the assessment of an object's 'worth' in terms of hard cash.

In the final analysis, however, such factors cancel out when large numbers of sales are taken into account by an experienced valuer, and it is possible to arrive at a surprisingly accurate assessment of current values of antiques; an assessment which may be taken confidently to be a fair indication of the worth of an object and which provides a reliable basis for negotiation.

Throughout this book, objects are grouped under category headings and, to expedite reference, they progress in price order within their own categories. Where the description states 'one of a pair' the value given is that for the pair sold as such.

1SBN 0 902921 48 7

CONTENTS

Pressed glass ale with simulated capstan stem, circa 1870. £4

U bowl ale with arch decoration, circa 1850. £7

U bowl ale glass with arch moulded decoration and knopped stem, circa 1840. £7

Short ale glass with wrythen decoration, 4½ ins. high, circa 1860. £9

U bowl ale with moulded cut thumbprint decoration, circa 1850. £9

Capstan stemmed ale glass with arch decoration, circa 1840. £10

U bowl ale glass engraved with hops and barley, circa 1840. £11

Short ale glass with lemon squeezer foot, circa 1800. £12

Late 18th century hop and barley engraved ale glass. £12

Late 18th century ribbed ale glass. £14

18th century ale glass with folded foot and trumpet shaped bowl etched with hops and barley design, circa 1790, 10ins. £35

Engraved ale or champagne flute, 1780, 19.7cm. high. £50

Cotton twist ale glass, circa 1770. £50

An airtwist ale glass, 7½ ins. high. £68

17th century ale glass with a wrythen bowl. £80

Airtwist ale glass, with round funnel bowl engraved with a single hop and two leaves, 1750, 19.7cm. £85

Mixed twist ale glass with a plain conical foot, 1770, 19.1cm. high. £90

Early English flute glass. £90

Opaque twist ale glass
with tall round funnel
bowl, 1760.   £95

English opaque
twist ale glass.
£100

Drawn trumpet bowl
ale glass on plain stem.
£130

Jacobite ale glass.
£170

An engraved Jacobite
ale glass with air twist
stem.   £175

Baluster stem ale
glass with triple
knop above an
inverted baluster,
circa 1740. £175

A double knopped air
twist ale glass, circa
1760.   £200

Engraved ale glass,
6,1/8 ins. high,
circa 1760. £210

A Beilby ale glass
decorated with barley
hops in white enamel.
£230

One of a pair of frosted glass bookends in the form of elephants, signed J. Hoffman, 14 cm. £26

Mid-19th century spun glass tableau of three peacocks, 47cm. £70

Pate de verre study of a stag beetle by Almeric Walter. £105

Pate de verre study of a blue green lizard by Almeric Walter. £145

Oval crystal quartz intaglio of an owl by Ronald Pennell, 18ct. gold. £190

A fine faience dog by Emille Galle, 12½ ins. 300 gns.

Faience cat by Emille Galle, 13 ins. 350 gns.

Faience cat by Galle, painted with a yellow body covered in blue hearts, circles and florettes, with blue ears and markings and green glass eyes, 12¾ ins. high. 380 gns.

Pate de verre chameleon paperweight by H. Berge in deep green glass, 1910, 8.6cm. high. £580

APOTHECARY BOXES

Mahogany apothecary box complete with bottles, circa 1820.   £38

Mahogany medicine case fitted with glass jars, mortar and pestle etc. 8¼ ins. wide.   £55

A mahogany apothecary cabinet with a secret poison compartment, twenty-six assorted bottles, and brass scales, circa 1790.   £60

19th century apothecary box complete with bottles.   £65

Late 18th century mahogany apothecary box complete with the original bottles. £70

Outstanding George III mahogany apothecary cabinet on bracket feet, with old bottles, pestle and mortar, 15ins., circa 1795.   £320

A Sowerby's of Gateshead 'blanc de lait', slag glass ornamental basket, 8 cm. wide. £8

19th century Cranberry glass basket with a plated stand. £18

Victorian silver sugar basket with a blue glass liner, London, 1846, 5 oz. £48

BEERSTEINS

An Imperial German cut glass half litre beerstein with a heavily plated lid surmounted by a Jaeger shako. £18

A fine Imperial German glass, half litre beerstein. £35

A good Imperial German Reservist's glass, half litre beerstein, painted with the Kaiser's portrait. £41

BISCUIT BARRELS

Victorian circular cut glass biscuit jar with a plated stand and cover. £13

Victorian frosted glass biscuit jar with a plated lid and handle. £15

Victorian engraved glass biscuit jar on a plated and engraved stand with bun feet. £18

BEAKERS

Late 19th century Venetian-style mauve glass beaker with white glass decoration. £8

Victorian glass beaker with wooden container. £10

Pink glass Mary Gregory beaker. £23

Lead glass beaker with an oval medallion, showing a square rigged sailing ship and 'Grove Hill', 6¾ ins. high. £37

A fine English beaker engraved with a church, a mill and a fort flying the Union Jack, circa 1760. £55

Late 18th century Bohemian double overlay beaker. £95

An early 18th century glass beaker with gilt decoration. £100

An opaque white glass beaker enamelled in colours, 9 cm. £290

18th century amber flask beaker by Kothgasser. £375

14

A beaker of grey Lithyalin glass by Freidrich Egermann of Blottendorf, 13ins. high, first half of 19th century. £480

Amber flask beaker by Kothgasser. £500

A fine amber coloured beaker signed Anton Kothgasser (1769-1851). £600

Landscape beaker from the Mohn workshop, circa 1815, 4½ in. £620

Lithyalin glass beaker by F. Egermann, 10.5cm. high, first half of 19th century. £780

A deep-coloured marbled glass beaker by F. Egermann, 4¼ in. high. £1,795

Early 19th century enamelled glass beaker decorated by Anton Kothgasser, 12cm. £1,995

Late 17th century wheel-engraved ruby glass beaker, South German, 12.5 cm. £2,500

Glass beaker with enamelled band of flowers by Mohn Jnr., 4½ in. £2,520

15

BOTTLES

Small green tinted
glass sauce bottle.
£0.75

Early light amber
glass 'Bovril' jar.
£1

Small mineral
water bottle.
£1

Elegant green glass
beer bottle, circa
1890.   £1

Sheared-top sauce
bottle of green
tinted glass.   £1

Brown glass beer
bottle.   £1

Cobalt blue poison
bottle.   £1.50

Amber glass hair
tonic bottle. £2

Bung-stoppered mineral
water bottle.   £2

Brown glass beer
bottle.   £2

Small green glass
poison bottle. £2

Cobalt blue bottle
marked 'Not to be
taken'.   £2

1 pint brown glass
beer bottle.   £3

German mineral water
bottle in dark green
glass.   £3

'Dumpy' green glass safecure
bottle for diabetics.   £4

Victorian brown glass
spirit bottle.   £4

Victorian whiskey bottle.
          £4

Amber glass poison
bottle.   £4

BOTTLES

Green glass 'onion' bottle, circa 1885.
£4

Clarke's clear fluid ammonia bottle.
£5

Dark green glass pickle jar 1880.
£6

Cobalt blue bottle marked 'Poison'.
£8

Codd's glass mineral bottle with amber stopper. £12

Cruet bottle with a geometric cut body, scale cut neck, and plated cap, circa 1810. £12

19th century green glass wine bottle. £12

Codd's 'light bulb' bottle with black marble stopper.
£15

Victorian embossed black glass whiskey bottle. £15

Opaque, deep blue glass
decanter bottle by
Mowart of Scotland,
circa 1920. £15

Mid-19th century Victorian
scent bottle. £17

Georgian green
glass wine
bottle. £17

Georgian wine
bottle, 10 in.
£18

Zara seal bottle. £20

Daffy's Elixir
Cure All bottle.
£28

Victorian decanter stand
with cut glass bottles.
£30

A blown and
moulded sealed
bottle dated
1802. £30

Price's Patent Candle Co.
cough medicine bottle
in cobalt blue glass. £30

BOTTLES

Codd's amber glass mineral bottle. £30

A Bristol blue glass ketchup bottle.   £35

Warner's green glass safecure bottle for diabetics.   £40

Rare bitters bottle dated 1874.
£45

A pair of small Bristol blue coloured sauce bottles with lozenge stoppers, 6 in. £50

An Oxford tavern bottle dated 1684.
£50

A blown and moulded sealed bottle dated 1775.
£60

19th century Nailsea bottle, 8in.   £60

An English sealed wine bottle, the seal showing a comet with five trailing rays, indicating the comet year 1811, 12¾ in. high.   £60

Egyptian grave-
ware tear bottle.
£60

Silver plated decanter
stand holding three
coloured glass bottles.
£65

Green glass
bottle, circa
1770. £65

19th century
satin glass
bottle, 7½
in. high. £75

H. Codd's cobalt
blue marble
stoppered bottle.
£100

Hand painted
Egyptian bottle,
400 BC. £160

18th century green
glass bottle, circa
1726.   £175

Rare Sidonian
bottle, circa
100 AD. £600

One of a pair of 19th
century French 'George
De Pigeon' opaline and
gilt bottles and stoppers,
9½ in. high.   £780

BOWLS

Orange carnival
glass bowl. £3

Victorian mauve carnival
glass bowl. £4.50

Small Victorian
amber glass
bowl. £5

Decorative Victorian
glass bowl. £7

Edwardian moulded
glass bowl with
plated rim. £7

Victorian ruby
glass bowl, 4 in.
diam. £7.50

Victorian heavy
cut glass pedestal
bowl. £10

Victorian circular
moulded glass bowl
on a plated stand,
7½ in. diam. £10

19th century
Bristol blue
finger bowl.
£10

BOWLS

Milk glass
crimped bowl.
£10

Edwardian green glass
dessert set comprising
six plates and a bowl,
25.5 cm. diam.   £16

White glass sugar
bowl, possibly
made on Tyneside.
£35

Late 19th century
Bohemian glass
bowl.   £35

French Lalique glass bowl
on four feet, 24cm. diam.
(chipped).   £38

Regency period glass
punchbowl, finely
engraved with grapes,
hops and barley, with
shield monogram H.M.M.
and dated 1806.   £58

Powder bowl by
Lalique decorated
with antelopes in
white glass, circa
1915.   £60

A Georgian cut glass
punchbowl and domed
cover, on cut stem and
square base, 17 in. £67

A Waterford glass
fruit bowl.   £80

23

BOWLS

Georgian cut glass punchbowl and cover, on stand, circa 1820. £107

Rare caddy bowl in cut blue glass, circa 1810. £110

Glass boat shaped bowl with bayeuge cutting. £140

A Galle cameo glass bowl, the pale pink body overlaid in green and etched with teasels, 19cm. wide. £150

Bristol finger bowl marked I. Jacobs, Bristol. £150

Glass bowl by Gabriel Argy Rousseau. £175

An early English glass covered bowl. £445

Fine quality bowl by Francois-Emile Decorchemont. £450

A fine English cameo-cut bowl. £1,000

19th century Nailsea candlestick. £37

Victorian cut glass candlestick. £4.50

A large ornate brass candlestick with crystal glass drops, 24½ in. high. £48

Art Deco dressing table set, circa 1930. £85

A pair of brass candelabra each for five lights with four branches, the bell shaped candleholders with wax pans and amber cut glass pear shaped drops, 23½ins. £175

One of a fine pair of French ormolu three branch candelabra with cut glass prism drops, 14 in. high, circa 1830. £175

A rare taperstick, the nozzle set on inverted baluster air twist stem, 6½ in. £370

Attractive old Sheffield plate centrepiece with cut glass bowl, circa 1850. £375

Glass candlestick with a stem comprising different varieties of knops, 7¼ in. £440

CAR MASCOTS

A fine Lalique glass car mascot in the form of a seagull in flight. £54

A Lalique car mascot in the form of a hawk in clear solid glass, signed on base. £130

Glass car mascot entitled 'Spirit of the Wind' by Rene Lalique, 10¼ in. long. £200

CENTREPIECES

Covered Bohemian comport, circa 1830. £28

19th century cut glass and plated centrepiece. £30

19th century cut glass and plated centrepiece. £30

One of a pair of French 19th century comports, with glass bowls raised on brass and silvered stems and bases, 9 in. £40

A five light Victorian plated centrepiece with glass dishes. £135

Three branch table centre in the form of an oak tree with stags, standing on mirror base, 27ins. £190

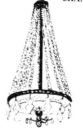

A French lantern-shaped electric
light, circa 1910, 97cm.   £60

Gilt metal and cut glass
electric chandelier.   £140

Early 19th century twelve
light chandelier with cut
glass drops.   £185

A 19th century brass and
glass chandelier with eight
scrolled arms.   £165

A glass chandelier with eight
fluted scroll branches. £200

Highly coloured Belgian
chandelier by Muller, 61 cm.
high.   £210

CHANDELIERS

A fine large early 19th century
chandelier with crystal glass
drops.   £250

George III cut glass chandelier
2 ft. diameter.   £275

A large Edwardian salon
chandelier with a fountain
of glass drops, 107 cm. £500

One of a pair of 19th
century cut glass chandeliers.
£820

A superb 18th century English
cut glass chandelier. £4,725

Superb Adam style chandelier,
circa 1785.   £6,000

Plain glass chemist's jar.
£1.50

Blue glass chemist's jar.
£2.50

Chemist's green glass
poison jar.   £3

Green glass carboy.
£7.50

A large globular purple
chemist's bottle with
stopper and gilt and black
"Syr. Phos. C." label,
14ins. high.   £30

A large apothecary's jar,
made of crude bottle
glass with enamel splatter.
£40

Victorian chemist's glass
jar with 'Rhubarb'
painted in gold and
blue, 40ins. high.   £48

Eglomisee glass chemist's
jar decorated with St.
George and the Dragon.
£155

Pair of apothecary jars,
marked 'Castor Oil
Seeds' and 'Cassia',
with original contents.
£195

29

CHEVAL MIRRORS

A fine mahogany framed cheval mirror, with shaped top and standing on cabriole legs. £28

19th century mahogany cheval mirror supported on paw feet. £35

A walnut framed cheval mirror with turned pillar supports. £38

Victorian mahogany cheval mirror on paw feet. £40

19th century birch upright cheval mirror on turned supports with brass candleholders. £45

19th century inlaid mahogany framed cheval mirror on fluted pillars with scroll feet. £60

A 19th century mahogany cheval mirror on fluted pillar supports. £65

Georgian cheval glass in a turned mahogany frame with brass candle holders. £90

George III mahogany cheval mirror, 5ft. 6 in. £110

Victorian cut glass
jug with plated
mounts.   £18

Late 19th century
plated claret jug.
                £22

Victorian glass
claret jug with
plated mounts.
                £28

Victorian clear
glass claret jug
with plated
mounts.   £30

An Edwardian silver
mounted claret jug.
                £45

George III claret
jug, circa 1820.
                £50

A baluster shaped glass
claret jug etched with
stylised acanthus leaves
and foliage banding,
12 in. high.   £52

Victorian red glass
claret jug with
plated mounts.
                £56

A tapered cylindrical
glass claret jug with
bulbous base and a
Victorian silver
mounted lid and
handle, 12 in.   £65

31

CLARET JUGS

Silver plated glass claret jug by C. Dresser.   £150

Victorian frosted glass claret jug, with silver mounts.   £161

Long necked silver mounted claret jug, made in London 1872 by E. C. Brown, 11½ in.   £295

A Guild of Handicrafts claret jug, with a green body in a silver frame, 8½in. high.   £330

Silver mounted claret jug, made by James Powell & Son, London, 1904.   £375

A French cameo glass claret jug by Daum, decorated with sprays of poppies and silver gilt mounts, 12½ in. high.   £500

Rare amber cameo claret jug with a plated mount. £550

A pair of 19th century claret jugs, the silver mounts representing the walrus and the carpenter, 15½ in. long and 8¾ in. high.   £650

Cameo glass claret jug by Webb.   £650

A convex mirror in carved gilt and ebonised frame, 22 in. diam.   £20

19th century circular gilt convex mirror.   £30

Regency period gilt convex mirror with eagle and scroll surmounts, 24½ in. diameter.   £65

Regency period convex mirror with the original gold leaf, circa 1820, 35 in. high, 22 in. wide.   £75

Early 19th century gilt convex mirror surmounted by an eagle.   £85

Regency water gilt convex mirror with original glass and in mint condition, circa 1810, overall height 30 in. diam. 20 in.   £150

COSMETIC BOXES

Early Victorian rosewood fitted vanity case.   £23

Lady's cosmetic box in rosewood with brass edging and stringing, containing seven pieces of glass and eight plated tops.   £30

Lady's rosewood, brass bound travelling toilet box, the glass jars and bottles with silver gilt lids, London, circa 1859.   £130

CRUETS

Victorian three bottle plated cruet (one missing).   £9

Late 19th century four bottle plated cruet.   £16

Victorian plated cruet stand complete with six cut glass bottles.   £25

Edwardian seven bottle plated cruet.   £25

Victorian six bottle plated cruet with cut glass bottles.   £35

A small four bottle plated cruet.   £35

Small silver novelty cruet in the form of a roller skate which carries the mark for 1837.   £35

A fine Victorian seven bottle cruet with cut glass bottles.   £40

Four bottle silver cruet, by R. Cattle and J. Barber, York, circa 1810.   £52

George III cruet stand by Robert Hennell, 1790. £120

Seven bottle silver condiment set, dated 1868. £120

George IV silver boat shaped cruet set and stand by H. M. London, circa 1822-23. £135

William IV silver cruet with cut glass bottles. £150

18th century six bottle Dublin cruet by G. Hill, 1766. £225

Georgian five bottle pierced silver cruet by T. Dealtry, London, 1776. £230

George II silver Warwick cruet with five glass bottles, circa 1760. £500

A good George III Warwick cruet. £630

A George III egg cruet for six, incorporating salt and a butter dish, 10 in. wide. £900

CUPS

A custard cup with cranberry coloured bowl and clear handle, stem and base. £5

A glass boot stirrup cup, circa 1800, 6½ in. high. £20

One of a pair of Chinese cherry red wine cups, 18th century, 3 in. high. £48

DECANTER BOXES

Small early 19th century mahogany decanter box containing four decanters. £45

A George III satinwood and kingwood banded decanter box, the lid and front inlaid with shields, 7 in. wide. £85

Georgian mahogany decanter set with six Venetian gilt decorated spirit decanters, circa 1795. £95

Coromandel and brass mounted liquor casket fitted with four cut glass decanters. £110

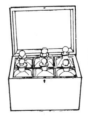

Set of six decanters, circa 1795. £120

An ebony decanter suite contained in a case inlaid with the Imperial Eagle and Bees. £355

Cut glass wine decanter and stopper, etched with vines. £8

19th century bun shaped decanter and stopper with slight disc cutting, 29cm. high. £4

Victorian decanter and stopper. £6

Late Victorian bulbous cut glass decanter, 8½ in. high. £12

Victorian hand-painted decanter. £15

A stirrup glass decanter, circa 1820. £15

George III decanter. £15

19th century cut crystal decanter of shouldered shape, 33 cm. high. £22

Plain glass decanter, circa 1800. £22

DECANTERS

One of a pair of cut glass square decanters and stoppers, 10ins.   £24

A small 19th century Bristol decanter in blue with a plated spout.   £25

George III square cut glass decanter.   £25

19th century Bristol blue decanter.   £28

A full size Georgian glass decanter.   £28

A green Bristol decanter.   £28

19th century green glass decanter.   £28.50

Victorian Mary Gregory dimpled decanter decorated with a white glass figure of a young girl.   £30

Mary Gregory green glass decanter.   £30

Early 19th century
cut glass decanter.
£34

Georgian engraved decanter,
circa 1780. £35

Bottle-sized glass
decanter. £35

Blue overlay decanter,
circa 1845. £36

A pale green Prussian style
decanter, with three-ring
neck and lozenge stopper,
circa 1840. £39

Bristol blue glass
decanter, circa
1800. £40

A Bristol amethyst
decanter with a
silver cork stopper.
£40

Rounded base decanter engraved
with floral festoons, with a snake
coiled round the neck, circa
1840. £45

Late 18th century
engraved decanter.
£45

DECANTERS

Rock crystal engraved
decanter, probably by
J. Keller, 10½ in. high,
circa 1885.   £48

Quart-sized decanter
with a central bank
of diamond shaped
cutting, circa 1810.
£50

18th century decanter
engraved 'Burgundy'.
£50

Ship's decanter,
circa 1850.
£60

18th century decanter
engraved 'Port'.   £60

One of a pair of
Georgian pint
size glass
decanters.   £72

Heavy ship's
decanter, circa
1860.   £75

A modernist cocktail service,
consisting of a decanter and
.six glasses, circa 1930.   £75

Adam style cut
and engraved
lipped and
shouldered
decanter, circa
1770.   £85

Sunderland Bridge engraved decanter, circa 1840.   £95

Quart-sized, wheel engraved decanter with thistles and roses and triple plain ringed neck, circa 1790.   £100

One of a pair of bulbous decanters with diamond cutting and double ringed neck, circa 1820.   £110

One of a pair of quart-sized decanters with bands of diamond and step cutting, circa 1820.   £110

One of a pair of decanters with broad bands of flute and diamond cutting, circa 1810.   £120

Ship's decanter with flute cutting and double ringed neck, circa 1810.   £150

One of a pair of Baccarat decanters with silver gilt caps, Paris 1819-38.   £200

Shouldered mallet shaped decanter, engraved 'Port', circa 1770.   £200

One of a set of four decanters with bands of diamond flute and step cutting, circa 1810.   £200

41

DECANTERS

A glass decanter by
Rene Lalique, 12ins.
high.   £200

Two of a set of four barrel
shaped blue spirit decanters
with gilt labels, circa 1790.
£250

One of a pair of ship's
decanters with flute
cutting and bull's eye
stoppers, circa 1810.
£250

An English glass
decanter, 10½ in.
high.   £680

A Bristol opaque white
decanter and stopper.
800 gns.

A Ravenscroft decanter
jug, circa 1685. £1,000

Late 17th century
English decanter
jug complete with
stopper. £1,225

A fine English decanter
and stopper of blue
glass with gilding,
27.3 cm.   £1,250

Silver mounted green glass
decanter by the Guild of
Handicrafts Ltd., 1901,
20.5 cm.   £1,500

A leaf shaped cranberry coloured lustred carnival glass dish, 8 in. diam. £6

Victorian white slag glass dish. £8

A cranberry coloured Murano type jam dish in an E.P.N.S. holder with spoon. £12

Glass butter dish cover and stand, circa 1810. £38

A Lalique glass circular dish, with groups of birds in relief, 8½ins. diam. £40

Mary Gregory amber glass pin tray depicting a boy with a butterfly net. £50

One of a pair of Victorian plated bird stands with glass dishes. £55

Mary Gregory enamelled dish with everted rim, circa 1880, 13¼ in. diam. £85

A high quality pate de cristal dish by Argy Rosseau. £190

43

DRESSING TABLE MIRRORS

A Victorian mahogany dressing glass, the rectangular plate raised on shaped and moulded supports on serpentine base, 1ft. 11ins. wide.   £7

Victorian walnut veneered dressing table mirror. £7

Late Victorian birchwood toilet mirror.   £10

Late Victorian, mahogany framed dressing table mirror.   £11

Victorian swing mirror with barley twist supports.   £11

19th century mahogany toilet mirror on a box base with drawer. £14

A mahogany dressing mirror with two drawers in the base, 2ft. wide.   £14

Early Victorian rosewood framed swing mirror with bobbin supports.   £15

A mahogany dressing glass with two drawers in the base, 2ft. 2in. wide.   £15

A Victorian walnut dressing glass, the oval plate raised in a 'U' shaped base with lidded compartments, 2ft. 1in. wide.   £15

19th century rosewood dressing table mirror with a drawer in the base.   £15

A mahogany framed oval dressing mirror.   £17

A mahogany dressing glass on box base with three drawers, 1ft. 10in. wide.   £18

A 19th century oval table mirror in brass frame, on oblong base.   £18

A mahogany framed oblong dressing mirror on oblong base with drawer, 15½ in. wide.   £20

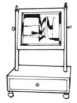

A small walnut inlaid dressing glass with turned spindle supports and stretcher, 12¼ in. high.   £20

19th century oblong mahogany dressing table mirror with drawer, 35 cm. wide.   £22

A 19th century tortoise-shell mounted, upright dressing glass, 16¾ in. high.   £25

DRESSING TABLE MIRRORS

A small mahogany inlaid oval dressing mirror, 15½ in. high.   £28

A Victorian mahogany and satinwood inlaid, upright dressing glass on bow front base with three short drawers, 1ft. 11in. wide.   £30

A mahogany upright dressing glass, raised on box base with three drawers, 15ins. wide.   £35

A mahogany inlaid dressing mirror on shaped box base with three drawers, 2ft. wide.   £35

An oval mahogany dressing mirror on scroll frame, 2ft. 4ins. high.   £35

A walnut upright dressing glass, the base with three drawers, 16 in. wide.   £35

William IV dressing glass with bevelled plate.   £36

19th century ebonised and gilt bamboo pattern, upright dressing mirror.   £39

A walnut upright dressing table mirror with shaped cornice on an oblong base with three drawers.   £40

A Regency mahogany toilet mirror, 31 in. high.   £42

Early 19th century serpentine front toilet mirror.   £45

Early 19th century Sheraton mahogany toilet mirror, the oval frame with satinwood stringing and brass handles at either side, 13 in. high.   £48

Regency mahogany rectangular toilet mirror.   £55

Late Georgian mahogany swing mirror.   £55

Papier mache toilet mirror, circa 1845.   £55

Late 18th century mahogany dressing table mirror with drawers in the base.   £60

A late 18th century, framed, oval dressing mirror, 2 ft. wide.   £60

Georgian mahogany toilet mirror, 19 in. wide, 26 in. high.   £62

DRESSING TABLE MIRRORS

Georgian vase shaped swing mirror. £65

A mahogany inlaid oval dressing glass on serpentine fronted base with three drawers, 21 in. high. £65

Georgian serpentine front mahogany swing mirror. £70

A late 18th century mahogany dressing glass with two drawers in the base, 23 in. wide. £70

Georgian mahogany toilet mirror with three drawers at the base supported on ogee feet. £85

George III mahogany swing mirror. £85

18th century mahogany swing mirror with shaped drawers in the base. £85

A 19th century rosewood inlaid dressing mirror with box base, 2 ft. wide. £85

Late 18th century mahogany toilet mirror. £95

Chippendale style toilet mirror in faded mahogany with the original mirror, circa 1760. £120

Mahogany inlaid mirror, three drawers in base, circa 1800. £130

18th century lacquered dressing table mirror with a fall front enclosing a fitted interior. £140

A Hepplewhite mahogany serpentine front box toilet mirror, 16¼ins. wide. £150

18th century Indian ivory and penwork mirror. £185

An early 18th century walnut and marquetry toilet mirror on bureau base. £220

Late Georgian mahogany swing mirror. £240

An early Victorian dressing table mirror with silver-gilt mounts, by Robert Hennell, 1843. £280

Walnut veneered toilet mirror with drawers in the base, circa 1725. £295

49

EPERGNES

Victorian cut glass centrepiece.
£27

Victorian spatter
glass epergne.
£15

Victorian cranberry glass
four branch epergne.
£28

Victorian white opaque
glass flower and fruit
epergne. £30

Victorian satin glass
fruit and flower
epergne. £40

Victorian crimson glass
epergne with centre
trumpet, 60cm. high
overall. £45

A Victorian silver gilt
epergne by Hunt &
Roskel. £380

George III silver gilt epergne
with glass dishes by M. T.,
London, 1806. £390

A parcel gilt epergne
by Elkington & Co.,
1875, 141 oz. £950

Heavy 19th century frosted and cut glass measuring ewer of superb quality, 14 in. high. £35

An early English glass ewer engraved with hops and barley, circa 1760. £58

Late 19th century glass ewer mounted in silver-gilt, 16 in. high. £170

Frosted glass wine ewer by C. and G. Fox 1856, with silver gilt mounts. £200

A rare 17th century diamond-point engraved ewer. £380

A Victorian red glass wine ewer. £800

FLASKS

Translucent Roman green glass flask. £16

Nailsea flask in the form of a pair of bellows. £35

19th century Nailsea flask. £35

51

FLASKS

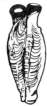

19th century Nailsea double flask with white loop decoration. £38

Pink, white and blue Nailsea flask, circa 1830. £38

Pink and white looped Nailsea flask, circa 1830. £38

Bohemian glass flask. £55

Scent flask decorated with gilt in the design of quiver and arrows. £110

Venetian 'Calcedonio' flask of shouldered hexagonal form, 17th century. £720

GIRANDOLES

19th century gesso girandole. £80

One of a pair of Italian giltwood girandoles, mid 18th century, 44in x 28in.£300

One of a pair of Irish glass girandoles from about 1785, 31½ in. high. 3,200 gns.

Tot glass with stylised
engraving.   £1

A large glass goblet,
the border etched
with thistles and
slice cut.   £20

Conical bowled goblet
engraved with a contin-
uous stag, circa 1730
8 in. high.   £22

Victorian red and
white Bohemian
glass goblet. £25

Crystal glass
goblet, circa
1800.   £40

Sunderland Bridge
goblet.   £48

One of a pair of English
goblets, finely engraved
and initialled W.D.,
circa 1800.   £55

Mixed twist wine goblet
with very large bowl,
1760, 19.1cm.   £100

An early goblet with
raspberry prunts and
a silver mount to the
foot dated 1924.£110

GOBLETS

A large goblet, circa 1700. £120

One of a pair of 1851 Great Exhibition goblets, both wheel engraved with scenes of Windsor Castle, 5¼ in. high.   £130

A goblet with a large straight sided bowl, honeycombe moulded, 9¼ in., circa 1760. £160

Large gilded goblet with ogee bowl on plain columnar stem and conical foot, 1750. 18.3cm.   £160

Lead glass goblet with a serpentine stem, circa 1685, 11½ in.   £220

A large lead glass goblet with prunts on the stem, 12½ins., circa 1790. £260

17th century engraved goblet from the Netherlands, 6 in. £290

Bohemian round funnel bowled goblet engraved with figures of a Boar Hunt circa 1740.   £290

Victorian overlay goblet 1850, cup-shaped bowl with crenallated rim, on a single knop and tall foot.   £380

English glass
goblet with
Jacobite
symbolism,
8½ins. £420

A Dutch engraved Newcastle
goblet with the arms of van
Brederode of Holland. £400

Elegant German goblet, about
1780, decorated with coloured
enamel. £400

A Dutch engraved Royal
Armorial goblet with the
Royal Arms of England,
20.5 cm. high. £425

Rare engraved Jacobite
goblet of seven petal
design and air twist
stem. £440

Venetian goblet of
the early 18th
century, 10½ in.
£480

English goblet with
a knop in the bowl
enclosing a coin
dated 1680, 17.2cm.
£700

A wheel-engraved
Silesian goblet, made
at Warmbrunn, circa
1760. £950

A goblet engraved with a
three-master within the
inscription 'Prosperity to
the East India Company,
Duke of Cumberland'.
£1,950

55

INKSTANDS

Victorian octagonal
glass ink bottle. £3

Square shaped plated
inkstand, with chased
border and glass ink-
well.   £8

Victorian brass
inkstand, circa
1850.   £8

Victorian brass
mounted inkstand
complete with
glass inkwell. £10

Edwardian silver ink-
stand, on four ball
feet with glass bottle,
4 oz. 15 dwt.   £13

Victorian glass and
marble inkstand.
                    £16

Travelling inkpot
by S. N., London,
1809, 4 cm. square.
                      £22

A Victorian brass inkstand
with two glass bottles. £22

A Victorian glass
inkwell with a
cat, circa 1860.
                    £23

INKSTANDS

Oblong silver inkstand with fluted border and glass ink-well, 6 in. wide, 3 oz. 10 dwt.   £28

Attractive Stourbridge inkwell decorated with millefiori design.   £65

A fine Stourbridge millefiori inkwell.   £75

Small George III inkstand by John Emes, circa 1798. £225

An ornate early Victorian double inkstand with two pots, central well and chamberstick with snuffer. £230

Art Nouveau glass inkwell. £350

Two bottle silver inkstand by Burrage Davenport, London 1777, 4 in. long 4 oz.   £425

Oblong silver inkstand on four panel feet by William Aboy, 17 oz. 10 dwt, 8¾ in.   £580

A rare Tiffany inkwell. £750

57

JUGS

A green glass jug with
waisted cylindrical
body and folded rim,
7½ in. high.   £3

A fine Victorian
Cranberry glass
water jug, 10in.
high.   £10

Glass jug with a
wrythen body
and applied
handle, circa
1800.   £10

A heavy cut glass
water jug, 11½ in.
high.   £10

Mary Gregory hand-
painted glass cream
jug.   £10

Victorian glass water
jug decorated with
embossed flowers. £12

19th century
Cranberry glass
jug.   £12

Purple slag
glass jug.
      £14

Victorian pink
glass jug with an
amber handle.
      £14

58

Victorian amber glass water jug. £15

Victorian ruby glass water jug, 10 in. high. £17

Victorian ruby glass water jug. £17

Victorian Cranberry glass jug, 14½cm. high. £17

A heavy George III plain glass water jug. £20

Good quality 19th century green and clear cut glass jug with a silver lip, 7 in. high. £21

Mary Gregory glass jug, 8 in. high. £24

Irish glass flat cut cream jug. £26

Georgian cut glass water jug. £27

19th century Mary Gregory decanter jug.   £28

A Lalique clear glass jug, the handle moulded with berries and foliage, 21cm. high.   £30

A small Georgian cut glass jug, 4½ in. high, circa 1810.   £35

English water jug, with broad flute cutting, about 1840, 8 in. high.   £39

Late Victorian ornamental jug, made at Stourbridge about 1880.   £44

A finely engraved and cut glass jug, 12 in. high, circa 1850.   £45

Irish glass yacht cream jug, circa 1800.   £45

A mid 19th century Bristol clear glass jug.   £45

Burmese glass jug, 4 in. high.   £45

Mid 19th century
Bristol clear glass
jug.   £45

English water jug with
cut swags, about 1820,
11 ins. high.   £60

A good quality 19th
century Loetz glass
jug.   £85

Roman jug, with bulbous
body, about 2nd century,
3 in. high.   £270

Venetian jug, about
1640, with dark tur-
quoise rim and
pinched lip.   £300

Irish water jug, about
1790, impressed Cork
Glass Co., round the
pontil mark, 6 in.
high.   £400

A Ravenscroft
crisselled decanter
jug with winged
ribs.   £625

A Ravenscroft syllabub
jug, gilt on sloping shoulders
with the label 'Honey
Syllabub'.   £2.500

Glass decanter jug
by G. Ravenscroft
with a heart-shaped
lid.   £3,400

LAMPS

Victorian green
glass night light.
£3.50

Victorian amber
glass night light.
£4

Victorian blue
glass night light.
£5

Brass circular oil lamp with
crimson glass shade, 20 in.
high.   £18

A brass oil lamp
with baluster stem,
white moulded
reservoir and shade.
£25

19th century blue and white
cut glass electric table lamp
and shade.   £25

Victorian style hanging
oil lamp.   £28

Unusual plated oil
lamp with a Nailsea
chimney.   £38

Brass hall lantern with four
leaded stained glass panels,
circa 1860, 22 in.   £45

62

A brass circular oil lamp with crimson glass shade, 23 in. high. £46

19th century plated electric table lamp with a cut glass shade. £55

A 19th century cut glass oil lamp, on circular base, with shade. £60

A lace maker's glass lamp, 16.5cm. high. £75

A delicate fuschia lamp by Muller Freres, circa 1905. £230

Art Deco table lamp by Daum of pink tinted opaque glass, 20 in. high. £230

Daum overlay lamp with purple and cloudy ground. £250

A Lalique table lamp in glass and iron. £300

Czechoslovakian art deco lamp, circa 1925. £450

63

LAMPS

A fine late 19th century lamp by Muller. £500

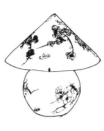

Galle glass table lamp, signed, about 1900, 35 cm. high. £530

One of a set of four Colza lamps on metal stands. £600

Tiffany lamp with a gilt bronze base and coloured glass poppy pattern shade. £820

A superb etched and enamelled glass table lamp. £920

Tiffany lamp. £1,750

Tiffany pendant coloured glass lightshade, 28 in. diam. £1,900

An excellent Tiffany lamp. £4,043

A good Tiffany 'Wisteria' lamp, 27 in. high. £6,500

Pair of Victorian ruby glass lustres.   £40

A pair of opaque glass lustres, with clear cut glass drops, 11 ins. high.   £50

A pair of rose lustres with original dumb-bell droppers, 15 ins. tall.   £55

A pair of 19th century overlay glass lustres, 12 in. high.   £75

A small pair of Victorian red and white overlaid glass lustres.   £80

Pair of ormolu and crystal lustre sidepieces, circa 1830.   £140

MEAD GLASSES

18th century mead glass. £48

Early 18th century mead glass. £175

A good early English glass. £175

Opaque twist mead glass with cup-shaped bowl incurved at the rim, 1750, 13.9cm. high. £200

A baluster mead glass with an incurved cup shaped bowl, circa 1710. £285

A 17th century English mead glass with a folded font. £315

MIRROR FIRESCREENS

Victorian brass and bevelled mirror firescreen. £10

Victorian brass firescreen, the mirror decorated with hand painted flowers. £14

Victorian mirror firescreen hand-painted with roses. £18

A Victorian silver
mounted mirror.
£15

Edwardian silver
mounted mirror.
£20

Victorian silver
framed standing
mirror, 10 in.
high. £22

An upright table
mirror in Art
Nouveau silver
frame. £34

Victorian silver
mounted table
mirror, 12 in.
high. £42

Charles II filigree-
fronted mirror,
1665, 9 cm.
across. £175

19th century Persian
mirror with a finely
painted lid signed
and dated 1840,
10¼ ins. x 7¼ ins.
£340

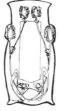

Silver and enamel
mirror by Liberty
& Co., 18¾ins.
£2,400

Silver gilt Louis XIV toilet
mirror, Paris 1660, (a
wedding present from
James II to the Duchess
of York). £50,000

MISCELLANEOUS

A large plated mirror plateau of rock design. £13

A large glass sphere with internal colourful transfers on a white ground, 12 in. diam. £22

One of a pair of Victorian amber glass 'domes of silence'. £5

Nailsea bell of red and blue mottled covering on a white ground. £75

Victorian stained glass leaded light window, 2ft. 4in. tall. £18

Late 19th century cut glass barrel with a gun metal tap. £22

A Georgian mother of pearl and silver, miniaturist's diminishing glass. £20

A gilded blue tulip design wine cooler signed by Isaac and Lazarus. £250

Palais Royal cut glass casket and cover with gilt metal mounts. £100

Mameluke enamelled glass
sweetmeat jar and cover.
£12,500

A large purple
hanging ball.
£15

Small ruby glass box
with metal mounts
and Mary Gregory
painting on lid. £28

Bohemian ruby glass
flagon with ormolu
mounts, 11 in. high.
£160

Late 19th century
hour glass egg timer
in Mauchline stand.
£8

An unusual bird drinking
fountain bearing 'The
Kingfisher', 17.8 cm.
£220

A carved cameo amber
glass plaque by Taid G.
Woodhall, 16.5 cm. high.
£1,500

Webb cameo glass
plaque carved by
H.J. Boam, 1885.
£5,800

Mid 19th century
Bristol clear glass
wine cooler with
characteristic
prismatic cutting
round the neck.
£40

MISCELLANEOUS

Victorian oil lamp
shade of pink glass.
£4

Victorian glass eyes. £6

A Lalique oyster shell
pattern ceiling bowl,
9½ in. diam.   £25

Glass walking stick. £20

A ruby glass desk rule
with silver mounts,
31 cm. long.   £20

Rare Lalique
panel initialled
L'oiseau Du
Feu, moulded
in relief, 17in.
high. £1,900

An illuminated Lalique glass
plaque carved with a naked
archer, 4ins. diam.   £400

Female figure in
glass by Lalique,
1925.   £185

Glass butter churn
mounted on an iron
stand.   £18

Bristol blue hunting
horn, circa 1820.
£34

A Daum match-holder of rectangular form, the pale blue frosted glass body enamelled with an Alpine scene, 4 cm. high. £42

Amber glass flower holder. £1.50

Glass toddy stick. £0.25

A fine gold leaf and enamel bevelled cut glass Heraldic shop sign, 12 in. diam, circa 1865. £58

A cut glass wine jug. £85

Bronze and glass standard lamp by Simon et Freres, 6ft. 8ins. high. £1,785

Victorian spun glass single masted fishing smack with a small blue rigged vessel, 36 cm. high. £75

Victorian glass ship under a glass dome. £80

Pair of figures by Lalique in frosted glass, 56 cm. high. £2,600

Victorian spun glass house, 51 cm. high, complete with glass dome and wooden stand. £100

71

OVERMANTELS

Late 19th century
mahogany overmantel.
£15

Overmantel mirror in
carved gilt frame of
Adam design, 4ft. 6ins.
wide.   £42

A giltwood overmantel
mirror, with arched and
moulded frame applied
with leaves and scrolls,
3ft. 6in. x 4ft. 5in. £50

A giltwood overmantel
mirror, of Regency
style, with three plates
and bordered by twist
mouldings, 2ft. 3ins. x
4ft. 5ins.   £55

Victorian overmantel mirror
in carved and gilt frame,
4ft. 9in. high.   £60

19th century gilt
and white Adam style
wall mirror with a
convex centre glass.
£120

A large late 19th
century gilt mirror
complete with seat,
10ft. 2in. high.£145

One of a pair of Chippendale
style overmantel mirrors,
circa 1760.   £600

An exceptionally fine
gilt overmantel mirror
in the manner of T.
Johnson, 6ft. 10in.
£1,900

Side view of Sentinel, an avant-garde paperweight. £12

Side view of Vortex by Colin Terris depicts a whirlpool movement. £20

Stourbridge millefiori paperweight. £44

Modern Scottish paperweight by Perthshire Paperweights Ltd. £45

Modern but very rare Scottish paperweight by Paul Ysart, signed PY. £200

Rare New England glass fruit weight, 3½ins. wide. £225

High quality St. Louis bouquet of fruit set in a basket of white latticinio. £240

American sandwich paperweight with one deep pink flower with pointed petals. £245

St. Louis magnum crown newel post weight, 5½ins. high. £300

73

PAPERWEIGHTS

Clichy paperweight with millefiori arrangement of two rows of florettes. £320

Close millefiori weight by Baccarat dated B1847. £385

Mid 19th century Clichy paperweight of a white dahlia with four leaves. £410

Fine St. Louis upright bouquet paperweight, 7.5cm. diam. £440

A Baccarat millefiori paperweight. £440

A St. Louis upright bouquet paperweight, 7.5cm. diam. £460

St. Louis fuchsia weight, 3¼ins. diam. £640

St. Louis 'crown' glass paperweight. £650

Clichy pansy flower delicately coloured with leaves of pale purple and lemon. £680

Baccarat glass paper-
weight, rare because
of its intricate work.
£950

Baccarat wheatflower
with pointed yellow
petals and black
markings.   £950

St. Louis fuchsia
paperweight. £1,000

Rare Baccarat paper-
weight showing a
butterfly hovering
over a flower. £1,250

Mid 19th century
French paperweight
by the Baccarat
factory.   £1,500

St. Louis green carpet
ground paperweight.
£1,600

Clichy glass bouquet
paperweight identifiable
by the pink ribbon
tying the stalks. £2,000

Bouquet paperweight
by Baccarat, superb
quality and condition.
£2,000

Clichy lily of the
valley weight.
£8,500

PIER GLASSES

Regency gilt and floral decorated pier mirror, 4ft. 2ins. tall, 2ft. 7ins. wide. £75

Victorian walnut framed pier glass with domed cresting, and inlaid overall with ebony and ivory, 6ft. 1in. high, 4ft. 1in. wide. £75

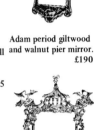

Adam period giltwood and walnut pier mirror. £190

Adam style pier glass surmounted by a lyre amid bocage and with acanthus capitals to the fluted pillars, 37ins. high, circa 1785 £200

Regency pier glass and table. £250

A fine Chippendale giltwood pier mirror in a delicately carved rococo frame. £300

Late 18th century Chippendale style carved giltwood pier glass. £500

William and Mary ebonised and silver gesso pier glass with silvered acanthus decoration with bead and reed moulding. £1,100

One of pair of pier glasses in the Chippendale manner. £2,500

PICKLE JARS

PICKLE JARS

Victorian glass pickle jar. £1.50

Edwardian glass pickle jar with floral engraving. £2.50

Pair of cut glass pickle jars and covers on a plated stand. £6

PIPES

Glass church warden pipe of clear glass with a blue bowl. £45

Pink and white Nailsea tobacco pipe, 41cm. long, circa 1830. £48

POTS

An iridescent glass honey-pot. £130

Silver mounted cameo and applied glass pot by Galle. £230

Galle glass perfume burner, 6½ins. high. £400

ROLLING PINS

Victorian glass rolling pin. £17

Bristol blue glass sailor's love token. £28

77

RUMMERS

A Georgian rummer
glass. £9

A clear Georgian
rummer. £11

Late 18th century rummer
engraved with a gentleman
fishing, 11.5cm. high. £16

Ale rummer engraved
with hops and barley,
circa 1880. £20

Engraved rummer of
traditional shape with
the initials S.J., 1800,
14cm. high. £30

Large ale rummer with
lemon squeezer foot,
circa 1810. £50

Sunderland Bridge
rummer, with the
initials G.A.H., 1800,
12.7cm. high. £60

Rummer with bucket bowl
engraved with a sailing ship
under the Sunderland Bridge
with the monogram M.J.G.,
5½ins. high. £65

Rummer engraved with
a view of St. Nicolas'
Church, Coventry, 1820,
15.2cm. high. £90

78

Victorian pierced silver
mustard pot, London
1890, 3ozs., with blue
glass liner. £13

Small, Irish glass
pair of salts. £16

Dutch silver mustard
pot with glass liner.
£20

One of a pair of Irish
step cut oval salts,
circa 1810. £25

One of a pair of George III
silver salts in the Adam
style with blue glass liners.
£25

One of a pair of Irish
salts, circa 1790. £34

One of a pair of Irish
glass boat shaped
salts, circa 1800. £40

George IV mustard pot
with a blue glass liner,
London. £90

One of a set of four William
IV chased and pierced salts
with blue glass liners,
Sheffield, 1828. £130

SCENT BOTTLES

Edwardian bottle
for lavender salts
with plated top.
£6

Porcelain perfume bottle
with silver cap, lead and
flower decoration in blue,
rust and yellow.   £8

19th century cut glass
scent bottle with a silver
hinged lid, 7ins. high. £14

Victorian green and
clear glass scent
bottle.   £16

Double ended blue
glass scent bottle
with silver gilt mounts.
£16

Victorian silver
mounted Bristol
blue scent bottle.
£16

Victorian vaseline
glass scent bottle.
£17

19th century green
scent bottle with a
silver screw top.
£18

Victorian ruby glass
double ended scent
bottle.   £18

Scent bottle by G. & W., Birmingham, 1899. £20

Green overlay scent bottle with silver mounts.   £18

Victorian blue and clear glass scent bottle.   £18

An Art Deco pink tinted, cut glass scent bottle. £24

19th century scent bottle in a pierced gilt metal mount inset with four landscape vignettes, 12cm. high. £25

19th century ruby glass scent bottle with a silver top. £25

Victorian egg-shaped scent bottle, silver mount and cap, by Mappin Bros., Birmingham, 1887.   £30

Mid 19th century gilt-topped, three bottle scent tantalus with leather case.   £30

Small Victorian gilt and enamel scent bottle. £30

81

SCENT BOTTLES

Victorian vaseline
glass scent bottle.
£30

Blue gilt and enamel
scent bottle.   £34

Fine double ended
blue overlay scent
bottle with silver
tops.   £35

A good 19th century
blue and white over-
lay scent bottle. £35

Mid 19th century
Bohemian glass
scent bottle. £40

A small early 18th
century glass scent
bottle.   £75

Bohemian glass
scent bottle of
three layers,
10in. high. £85

Bohemian scent
bottle with pain-
ted overlay and
gilt decoration,
19cm. high. £90

Victorian overlay
scent bottle. £95

Scent flask and smelling salts bottle by Sampson Mordan & Co., circa 1880, 7cm. high. £140

A large early 18th century gilded glass scent bottle. £150

Early 18th century glass scent bottle decorated with a garden scene. £150

Cameo glass scent bottle 6ins. high, fitted with a silver top. £185

Dutch glass scent bottle. £350

Scent bottle by Tiffany. £500

Two amber ground cameo bottles, made by Thomas Webb, or Stevens & Williams. £780

Elegant scent bottle possibly decorated in London. £892

Clichy patterned millefiori scent bottle and stopper, 16.5cm. high. £945

SNUFF BOTTLES

19th century glass interior painted snuff bottle. £30

19th century interior painted snuff bottle. £45

A 19th century white glass snuff bottle with red overlay. £50

Interior painted snuff bottle. £70

Peking snuff bottle overlaid with red flowers. £120

Chinese overlay snuff bottle. £130

Early interior painted Chinese snuff bottle. £160

Unusual interior painted snuff bottle. £180

A Chinese glass snuff bottle painted on the inside with horses. £210

Chinese overlay
snuff bottle.
£220

Early Chinese
snuff bottle.
£250

A fine 18th century
red overlay snuff
bottle.   £300

An opaque turquoise snuff
bottle, the white and black
double overlays carved in
high relief, 2½ins. high.
£315

Disc-shaped opaque white
bottle, brilliantly enamelled
in colours in the Ku Yueh
Hsuan style, 2¼ins. high.
£400

A Ch'ien Lung Peking
enamel snuff bottle
painted in famille rose
colours.   £1,050

A rare portrait snuff
bottle by Tzu I-tzu,
the reverse side bearing
an inscription. £1,200

A very rare interior-painted
glass snuff bottle of
flattened upright form,
signed T'ing Yu-Keng,
dated 'Winter month, 1904'.
£1,800

Important Chinese snuff
bottle in Canton enamel
with a gilt chased metal
stopper, blue Ch'ien
mark, 2¼ins. high. £4,000

SUGAR CASTERS

Edwardian glass sugar caster with a plated lid.   £5

Edwardian glass sugar caster with silver top.   £10

Victorian ruby glass sugar sifter, with silver top.   £12

TANKARDS

Pressed glass ale can with geometric decoration, circa 1860.   £7

A beer mug of bottle glass with white enamel splatter and a fine early handle.   £35

19th century Mary Gregory sapphire blue tankard, 6½ins. tall.   £58

Rare 18th century child's glass ale mug finely engraved with hops and barley motif and the inscription 'Cathleene', 4½ins. high.   £68

Bohemian Milchglas mug decorated with a roundel enclosing a figure in coloured enamel, about 1750.   £250

Mid 18th century engraved glass tankard.   £300

Oak tantalus spirit frame with two glass decanters and stoppers. £34

Victorian oak tantalus spirit case with three moulded glass decanters and stoppers. £35

Victorian three bottle oak cased tantalus with brass mounts. £55

A Victorian silver plated tantalus of three cut glass decanters. £60

A mahogany tantalus spirit frame with plated handles and mounts, and three cut glass decanters. £80

A fine Victorian three bottle tantalus in an oak case with silver mounts. £90

TAZZA

Late 19th century engraved glass tazza, 7½ins. diam. £12

17th century Latticino tazza, 8in. diam. £200

Empire turquoise verre opaline mounted tazza. £270

TUMBLERS

Newcastle slag glass tumbler in purple and white. £4.50

'Last drop' ale tumbler, circa 1870. £25

Farmer's tumbler with agricultural emblems and the motto 'Speed the Plough', circa 1820. £32

Georgian tumbler inscribed 'The Independence of Durham' and 'Rich'd Wharton its Defender', 4ins. high. £35

Georgian tumbler with a pale blue rim inscribed Mary Fogg, with figures of Faith, Hope and Charity, 3½ins. high. £45

Georgian tumbler engraved 'Success to Change Ringing', 4ins. high. £55

A Mildner tumbler. £800

A rare Jacobite portrait tumbler with a portrait of the Young Pretender. £800

A rare early Jacobite portrait tumbler engraved with a bust portrait of the Young Pretender. £850

A clear Victorian cylindrical twist bodied vase with flared neck.   £3

Victorian opal glass vase, 3½ins. high.   £3

Orange Carnival glass vase, 12ins. high.   £3

Victorian glass spill vase, 7ins. high.   £3

Victorian red glass spill vase.   £3.50

Slim Art Nouveau vase in pale blue glass.   £5

Iridescent green blue vase with tear drop decoration.   £6

Three handled Art Nouveau glass vase in green.   £8

Multi-coloured Victorian glass vase, 6ins. high. £8

VASES

One of a pair of
Victorian opal
glass vases, 4ins.
high.   £8

Victorian opal
glass vase, 10ins.
high.   £10

A tall Bohemian green
glass vase, with oval
opaque panel painted
with three figures of
cupids, 19½ins. high.
£10

One of a pair of late
Victorian cut glass
vases, 9½ins. tall.
£10

One of a pair of
iridescent carnival
glass vases, circa
1890, 10½ins. high.
£12

Very decorative Bohemian
glass vase with overlay white
on cranberry background,
on 5in. circular base, circa
1845.   £12

Victorian ruby
glass vase. £12

An attractive Victorian
vaseline glass vase with
opaque frilly edges on
top and bottom rims,
circa 1860, 6ins. high. £14

19th century
decorated glass
vase, 11ins. high.
£14

German clear glass
vase in a silver
plated pewter case
by Orivit. £15

Edwardian celery vase
engraved with ferns,
26cm. high. £16

Victorian pink satin
glass vase, 7ins. high.
£16

One of a pair of
Victorian opal
glass vases, 12ins.
high. £16

19th century Bohemian
gilded glass vase £17

One of a pair of Victorian
coloured glass vases
decorated with flowers,
10ins. high. £20

Satin glass vase
with gilt floral
decoration,
5ins. high. £20

19th century baluster
vase in iridescent mauve
and green. £20

Dark green Art Nouveau
rippled glass vase in a
brass case with lily pad
design. £22

VASES

Bubble glass vase painted with a 19th century military scene. £23

Iridescent green and yellow **Loetz** glass vase with applied snake decoration. £23

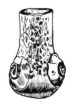

Art Nouveau vase decorated with snails in fiery orange and silver blue. £26

Moser iridescent glass vase in graded mulberry tinged with pink. £28

One of a pair of Mary Gregory cranberry glass vases, 5ins. tall. £28

19th century Bohemian red and white overlay glass vase. £30

One of a pair of 19th century ruby and decorated glass vases. £30

Mary Gregory pink glass vase of globular shape. £30

A balustroid vase in iridescent blue green, 7¾ins. high. £35

92

One of a pair of Victorian opal glass vases, 10ins. high. £35

Green glass tulip vase with blood red at the tips of the petals. £38

Art Nouveau glass vase in pale lime green with brown bullrushes rising from the base. £40

One of a pair of Mary Gregory cranberry glass vases, each decorated with the figure of a girl in white enamel. £40

Art Nouveau glass vase. £50

A fluted glass lily vase, 38ins. high, circa 1870. £55

Bohemian vase overlaid with white panels and decorated with gilt, 5.5in. high. £60

A Galle cameo glass vase of uniform shape, the grey orange body overlaid in orange, 9cm. high. £60

Green glass Bohemian vase with portrait on panel and gilt decoration, 19cm. £65

93

VASES

Large Mary Gregory
vase of pink glass.
£68

Clutha glass vase in
olive green with white
and blue striations,
circa 1895.   £75

A Galle cameo small
baluster vase, the
olive green body over-
laid in darker green.
£78

One of a pair of opaline
overlay pink vases, with
gold arabesques and
polychrome floral
decorations, 11 ins. high.
£85

A Legras cameo glass
vase of quatrefoil shape,
the frosted glass body
overlaid in purple,
13cm. high.   £85

An iridescent Loetz
glass vase, decorated
with waved lines on
a gold ground, 6ins.
high.   £95

Miniature cameo
vase, Galle.   £95

One of a pair of 18th
century Venetian
vases.   £120

A Pilkington Royal
Lancastrian vase
designed by Walter
Crane.   £120

VASES

A fine Galle vase,
signed. £135

Richardsen over-
laid glass vase,
11in. high. £135

Red glass Bohemian
vase with gilt deco-
ration and painted
panels, 19cm. £150

A Loetz iridescent
vase decorated with
green and blue.
£160

Glass vase from the
Daum workshop
decorated with a
raised design of
mushrooms. £160

A Webb yellow
and white satin
glass vase. £175

A Lalique vase decorated
with a raised design of
small scaly fish, signed
R. Lalique, France,
9ins. high. £175

Ch'ien Lung
overlay vase.
£175

A fine trial vase
by Rene Lalique.
£180

VASES

A Pilkington lustre vase painted with a continuous pattern of rampant griffins, scrolls and flowers in cobalt blue and gold. £185

A Loetz peacock blue iridescent glass vase, 25.5cm. high. £190

Tiffany amber lustre vase. £200

A Galle, green glass trumpet vase, 16ins. high. £200

A Galle glass acid etched cameo vase. £200

An Orrefors glass vase, by Vicke Lindstrand, circa 1930, 3½ins. high. £210

A Webb glass cameo vase, 9ins. high. £210

A Galle cameo glass vase of tall slender baluster form, the grey orange body overlaid in crimson, 35cm. high. £220

Tiffany peacock lily-pad vase, 1900. £230

Iridescent Loetz blue vase decorated with wavy lines. £230

19th century Tiffany iridescent golden vase, 10ins. high. £250

A Galle cameo glass vase overlaid in brown, 32cm. high. £260

A fine blue and amber Loetz glass vase. £260

Cameo footed vase set with four cabochons by Daum, circa 1905. £285

Blue and white carved cameo glass vase, 20.5cm. high. £340

A deep etched glass vase by Daum, circa 1925. £360

A Loetz glass bottle vase with a blue base decorated with flame designs. £360

A fine white and blue Webb cameo vase. £375

VASES

Galle cameo glass vase, overlaid in brown, 43.5cm. high. £400

Angular pate de cristal vase by Gabriel Argy Rousseau, cast in emerald green marbled glass, 1925. £400

One of a pair of Bohemian green and gilt vases, 17ins. high. £400

A cameo glass vase of smokey quartz colour, 9.5cm. high. £400

Burgun and Schwerer wheel carved vase in bluebell purple and gilt. £400

A good heavy art deco glass vase, by Andre Thuret, 1930. £420

A pate de crystal vase by Francois Emille Decorchemont, circa 1927. £450

A Webb white and blue cameo vase, 6ins. tall. £450

A superb vase by Rene Lalique, circa 1925. £480

Raisin-coloured
cameo vase,
30.5cm. high.
£546

A Tiffany Cypriot
vase.   £550

Tiffany feathered vase
richly feathered in
green, gold and peacock,
iridescent shades, 1900.
£700

Deep amber glass
serpent vase by
Rene Lalique, 24ins.
high.   £850

One of a pair of opaline
glass vases with flowers
in colours on a turquoise
ground, 39.5cm. high.
£850

A Galle cameo and
marquetry glass vase
with silver mounts
by Cardeilhac. £925

A superb Thomas
Webb vase. £925

Tiffany Lava/Cypriot
glass vase, in pale yellow,
1900.   £1,600

A Victorian four-
colour oviform
vase.   £1,650

VASES

A superb Lalique vase with a border of bacchantes. £1,800

Gold and yellow iridescent vase by Louis C. Tiffany, 38cm. high, circa 1900. £2,000

Cameo glass vase by George Woodhall, 8ins. high. £2,000

One of a fine pair of French mid 19th century Cristalleries de St. Louis vases with 'crown' paperweight bases, 25.7cm. high. £2,300

Argy-Rousseau pate-de-verre vase, 12¼ins. high. £2,500

A fine and rare marquetry vase by Emile Galle, made in 1900. £2,500

Pate-de-cristal vase by Francois Decorchemont in deep blue glass, 25.5cm. high, circa 1910. £3,000

A rare Jack-in-the-pulpit Tiffany peacock iridescent glass vase, 1900. £3,600

George Woodhall cameo glass vase, 12ins. high, with kingfisher blue ground. £8,925

Late 19th century ebonised frame hanging mirror. £14

Victorian upright wall mirror, 2ft. 5ins. high. £22

A walnut framed oblong wall mirror with scroll shaped cornice, 3ft. wide, 2ft 2ins. high. £22

19th century brass framed wall mirror, 15ins. high. £25

19th century bevelled wall mirror with side plates in embossed brass, 23ins. high. £25

19th century bevelled wall mirror in mahogany frame, 3ft. 3ins. high. £30

19th century oval gilt framed and bevelled wall mirror with ribbon surmounts. £35

A carved gilt wall bracket with three shelves and mirror to back, 3ft. 6ins. high. £36

An upright wall mirror in gilt fluted pillar frame with blue and gilt panel, 25¼ins. x 14ins. £38

WALL MIRRORS

19th century wall mirror in carved gilt frame, 2ft. 10ins. tall. £40

An oval bevelled wall mirror in gilt fluted frame with ribbon cresting, 3ft. high. £40

19th century gilded wall mirror, 3ft. high. £40

19th century upright wall mirror in shaped mahogany and gilt frame, 2ft. 10ins. high. £40

A Dutch upright wall mirror in elaborately carved oak frame, with female and cherub heads, and scrolls, 2ft. 6ins. x 2ft. 2ins. £42

An upright wall mirror in mahogany frame, the shaped surmount with a gilt figure of an eagle, 2ft. 7ins. high. £42

A reproduction yew wood and carved gilt framed upright wall mirror, the cresting with a gilt bird, 3ft. high. £42

19th century mahogany bevelled wall mirror, 83cm. wide. £42

An upright wall mirror in ebonised and carved gilt frame, with three candle branches at base, 2ft.9ins. high. £42

102

An oval shaped wall mirror in Venetian carved giltwood frame of scroll design with pierced surmount, 3ft. 8ins. high. £45

Victorian gilded mirror with gesso ornamentation. £45

19th century oval wall mirror in Italian carved gilt frame with four scroll candle sconces, 4ft. 6ins. high. £48

George III walnut framed upright bevelled wall mirror, 84cm. high. £50

Victorian oval gilt plaster wall mirror, 62ins. high, 32ins. wide. £50

One of a pair of scroll shaped wall mirrors in carved gilt frames, with three candle branches, 2ft. 9ins. high. £50

An upright wall mirror in a mahogany frame, the scroll cresting with carved giltwood cartouche, 3ft. 5ins. high x 1ft. 10ins. £60

Art Nouveau composite mirror in a wooden frame with applied chased copper panels. £60

Mahogany wall mirror with scroll mounts, 3ft. x 1ft. 9ins. £65

103

WALL MIRRORS

A 19th century bevelled mirror with three plates, in giltwood frame, 4ft. 2ins. x 1ft. 7ins. £65

19th century gesso wall mirror in a gilded frame. £65

19th century reproduction mahogany framed wall mirror surmounted by a gilt bird. £75

19th century Continental carved walnut mirror, 26ins. x 24ins. £75

Queen Anne style carved giltwood upright mirror, 89cm. high. £75

19th century Venetian carved and pierced gilt framed oval wall mirror, 82cm. £75

A large 19th century wall mirror in an elaborately carved frame, 176cm. wide. £80

19th century shield shaped scroll wall mirror of carved gilt wood, circa 1820, 26in. high, 17in. wide. £90

Mid 18th century mahogany fret mirror with carved and gilded Ho Ho bird, circa 1760. £125

Early 19th century
French brass mirror.
£125

A fine Georgian period
Irish carved giltwood
mirror.   £150

18th century mirror
by C. Carton.   £185

William IV carved wood
and gilded mirror,
34ins. x 52ins.   £185

Good quality George III
carved giltwood mirror.
£200

Late 18th century gilt
mirror with the original
glass and Verre Eglomisee
panel.   £240

Dresden china mirror,
circa 1860, 2ft. 8ins.
high.   £340

Early 18th century
mahogany and parcel-
gilt wall mirror, 44ins.
high, 27ins. wide. £340

One of a pair of late
18th century carved
giltwood mirrors.
£350

WALL MIRRORS

18th century giltwood mirror with superb deep carving.   £365

19th century Chippendale style carved mahogany mirror.   £375

A George I period giltwood and gesso wall mirror.   £400

George III giltwood mirror, with scroll pediment, overall measurements, 4ft. x 2ft. 1½ins.   £450

George II carved giltwood mirror of excellent proportions.   £485

Late 18th century Chippendale gilt mirror (in need of restoration).   £500

Late 18th century Chippendale style carved giltwood mirror.   £800

Chippendale period carved and giltwood mirror, 49ins. x 27ins.   £820

One of a pair of early 19th century Austrian carved and gilded mirrors in superb original condition.   £900

A George II walnut and gilt gesso mirror. £1,000

A superb early 18th century walnut and gilt mirror, surmounted by an eagle with outstretched wings. £1,000

A good early walnut and mahogany framed mirror, 54ins. x 30ins. £1,250

George I hanging mirror in a shaped brass frame with carved giltwood borders. £1,300

Mirror surrounded by stumpwork embroidery showing King Charles II and Catherine of Braganza. £1,300

An exceptionally fine Chinese Chippendale wall mirror, 3ft tall, 2ft. 6ins. wide. £1,500

One of a pair of 18th century giltwood mirrors, 48ins. high. £2,250

Early 18th century English upright mirror by William Kent. £2,500

English wall mirror, the frame decorated with japanning and paper filigree work, 127cm. high, circa 1700. £5,200

WINE GLASSES

Victorian ruby wine glass. £3

Bohemian ruby and engraved grapevine drinking glass. £6

Deceptive glass with capstan stem, circa 1810: £6

Engraved ruby sherry glass, circa 1825. £7.50

Multiple air twist wine glass (chipped base). £15

Gin glass with thick ribbed bowl, circa 1780. £15

One of a set of six Edwardian champagne glasses. £17

Mary Gregory glass. £17

Dram glass with a round bowl engraved with a fruiting vine and bird, circa 1750. £17

Wine glass with round bowl and moulded decoration, circa 1750. £17

Facet stem wine glass, circa 1790. £20

A wine glass, the waisted ogee bowl spirally moulded to half height, supported on an opaque corkscrew stem, 6ins. high. £21

Dram or spirit glass with moulded cup bowl on a plain stem, 1780, 9.9cm. high. £25

A wine glass with half moulded funnel bowl, on a multi-spiral and gauze opaque twist stem. £25

Facet stemmed wine glass. £25

Rare opaque twist wine glass with drawn trumpet bowl, 1770, 17.1cm. high. £25

18th century wine glass, the moulded bowl with engraved border, 5¾ins. high. £25

18th century cotton twist stem glass. £28

WINE GLASSES

Wine glass with funnel bowl on opaque twist stem and plain foot, circa 1760.   £30

Late 18th century air twist stem glass.   £30

An opaque twist stem wine glass, circa 1765. £30

18th century wine glass with diamond stem and engraved bowl.   £30

Coaching glass with faceted ball knop joined to an ogee bowl by a collar, 1820, 12.6cm. high.   £30

An unusual amethyst glass, circa 1850. £30

Soda glass with bell bowl on air twist stem and plain foot, circa 1760.   £32

An opaque twist wine glass. £34

Facet stem wine glass with round ogee bowl, diamond cut stem and plain foot, 1780, 15.8cm. high.   £35

Multiple series air
twist wine glass,
circa 1760.   £35

Plain firing glass with
trumpet bowl set on
thick flat foot, 1740,
8.8cm. high.   £35

Opaque twist wine
glass with deep
ogee bowl, 1770,
16.2cm. high.   £35

An opaque twist
stem wine glass,
circa 1760.   £35

An opaque twist
stem wine glass,
circa 1760.   £36

A plain stem wine
glass with bulbous
knop and ogee bowl,
circa 1750.   £37

Engraved English
wine glass, 5¾ins.
high.   £37

A plain stem firing
glass with terraced
foot, circa 1750.
                    £38

Early English
drinking glass.
        £38

WINE GLASSES

An opaque twist glass. £38

18th century colour twist green and red wine glass. £40

English trumpet bowl wine glass, 6¼ins. high. £40

Drawn trumpet bowl glass with double series opaque twist stem, circa 1760. £40

Faceted champagne flute with plain round funnel bowl engraved round the rim, 1790, 17.1cm. high. £40

18th century wine glass with an engraved bowl. £40

An opaque twist wine glass with floral decoration to the bowl. £40

Opaque twist wine glass with funnel bowl, 1760, 15.4cm. high. £40

Opaque twist wine glass with flared funnel bowl set on a double series opaque twist stem on a high conical plain foot, 1770, 13.8cm. high. £40

112

Opaque twist wine glass with ogee bowl vertically ladder moulded for its whole height, 1770, 15cm. high. £40

Opaque twist firing glass set on double series twist stem, 1770, 10cm. high. £40

Opaque twist wine glass with medium-size bucket bowl, 1760, 16.5cm. high. £40

Early English water flute wine glass, 6¼ins. high. £45

18th century wine glass with funnel bowl and air twist stem. £45

One of a pair of early wine glasses with an air twist stem. £45

Opaque twist wine glass with pan top bowl flared at the rim, 1770, 14.6cm. high. £45

An early balustroid wine glass. £48

Faceted wine glass with slightly flared ogee bowl and a petal cut and scalloped foot, 1770, 15.2cm. high. £50

113

WINE GLASSES

Wine glass with funnel bowl, vertically moulded and ladder ribbed to its full height, 1770, 15.2cm. high. £50

Drawn stem English wine glass, 7ins. high. £50

Wine glass with flared 'tulip' bowl, 1760, 16.5cm. high. £55

An excise wine glass with a round funnel bowl engraved with a sprig of fruiting vine, 6¼ins. high, circa 1745. £60

19th century wine glass. £58

18th century wine glass, 3,7/8ins. high. £65

Early engraved wine glass. £65

Baluster stem wine glass with single tear. £70

Cordial wine glass with small straight sided bowl, 1770, 13.9cm. high. £75

Early kit-kat glass
with a single tear
on the stem. £75

Victorian Bohemian
'documentary' stained-
glass goblet, engraved
on one side, with faceted
stem and circular foot,
1851.   £80

Cordial glass with a
flared bucket bowl,
6¾ins. high, circa
1760.   £80

Air twist cordial glass with
fine drawn trumpet bowl
set on a multi-ply mercury
air twist stem, 1745, 14cm.
high.   £80

Engraved cordial glass
with trumpet bowl
set on plain drawn stem
and conical foot, 1745,
17.8cm. high.   £80

Wine glass with a
bell bowl set above
a mixed twist stem,
circa 1760.   £80

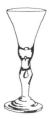

Wine glass vertically moulded
funnel bowl decorated with
eight panels, 1750, 15.8cm.
high.   £85

Engraved multi-spiral
air twist stemmed
wine glass, 6ins. high,
circa 1760.   £90

18th century
wine glass. £90

Engraved mixed twist wine glass, 1760, 15.2cm. high. £90

Cordial glass with bowl engraved with flowers and leaves, 1780, 16.5cm. high. £90

Mixed twist wine glass with a waisted bell bowl, 1755, 18.7cm. high. £90

Mixed twist wine glass with round funnel bowl set on rare stem 1760, 14.6cm. high. £90

Engraved cordial glass with small bowl moulded to half height, 1770. 17.1cm. high. £90

Opaque twist wine glass with plain ogee bowl set on vertically drawn opaque stem, 1760, 15.8cm. high. £90

Mid 18th century opaque twist wine glass. £90

A composite wine glass with a waisted bowl set over a multiple spiral air twist section terminating in a squat beaded, inverted baluster knop, 7ins. high, circa 1745. £100

18th century wine glass. £100

Jacobite wine glass engraved with a rose spray and two buds. £110

Opaque twist wine glass with octagonal moulded bowl, 1770, 15.2cm. high. £110

Colour twist wine glass with opaque white spiral gauze and plain opaque white spiral edged in translucent blue, 5¾ins. high. £110

Lynn wine glass, the ogee bowl with five horizontal rings, 1760, 13.6cm. high. £110

Mid 18th century air twist wine glass. 110gns.

Ratafia glass with a narrow straight sided funnel bowl moulded to two thirds of its height, circa 1745. £115

Wine glass with mixed twist stem, 7.5/8ins. high, circa 1760. £115

English opaque twist wine glass. £125

Light baluster stem wine glass with a domed and folded foot. £125

WINE GLASSES

'Captain' glass with large ogee bowl supported on a heavy double series opaque stem, 1760, 17.2cm. high. £130

Incised twist bright emerald green wine glass on incised stem and foot to match, 1750, 13.3cm. high. £130

A triple knop solid base baluster wine glass. £135

A Beilby ogee bowl wine glass 5¼ in. high. £145

18th century wine glass. £145

One of a pair of mid 18th century glasses. £150

A wine glass engraved with flowers, apples and pears. £160

Beilby wine glass with conical bowl enamelled in white, 1780, 15.2 cm. high. £160

An early English wine glass. £170

Opaque twist wine glass, gilded, with ogee bowl, 1770, 14.6 cm.   £175

18th century wine glass, 8.1/8 in. high.   £175

An ogee bowl wine glass with colour twist stem, 5¾ in. circa 1770.   £180

A colour twist wine glass with a bell bowl, 6¾ in. high.   £190

Rare bright olive green wine glass, the rim of which is heavily gilded, 1750, 16.5cm.   £200

Williamite wine glass.   £200

An engraved colour twist wine glass with fruiting vine and bird on a white lace twist stem surrounded by two translucent green spirals, 7¼ in.   £200

Newcastle glass with a large slightly flared round funnel bowl, 9 in. high. circa 1745.£210

Beilby wine glass with ogee bowl enamelled in white with a band of baroque rococo scrolling, 1770, 15.2 cm.   £210

119

WINE GLASSES

Baluster wine glass.   £225

Heavy baluster stem wine glass with tear in the knop.   £230

A baluster wine glass with a round funnel bowl, 5¾ins. high, circa 1710.   £265

Newcastle wine glass engraved in Holland.   £270

German white glass enamelled in colour, 15cm. high.   £270

Dutch engraved Newcastle glass.   £275

An early wine glass, the waisted trumpet bowl set in a stem of coloured and opaque twist.   £300

A magnificent wine glass with a conical bowl, solid at the base and set on a four sided Silesian stem, circa 1715.   £300

Early 18th century wine glass with single tear.   £320

Twist stem Jacobite
wine glass.   £350

A rare Beilby
enamelled
Masonic glass.
£400

A rare pan topped colour
twist wine glass, the stem
containing a central red
brick thread surrounded
by two opaque white
spirals, 5¾ins. high. £400

Beilby enamelled wine
glass with gilded rim,
1770, 17.4cm. high.
£425

A fine Newcastle-type
glass, engraved in
Holland.   £420

A Beilby wine glass with
a flared round funnel
bowl decorated with a
landscape scene in
coloured enamel. £430

A superb 17th century
baluster stem wine
glass.   £475

An 18th century coin glass,
with a bell-shaped bowl above
a hollow knop containing a
George III threepenny piece,
circa 1762, 6¾ins. high. £480

A rare Williamite
glass.   £500

WINE GLASSES

One of a pair of colour twist wine glasses. £850

An 18th century Amen glass with the crowned monogram of the Pretender James III of England and VIII of Scotland and underneath the word Amen, (repaired foot). £880

An English privateer glass inscribed 'Success to Oliver Cromwell, Paul Flyn, Commander'. £1,100

Stipple wine glass with decoration in the style of David Wolff, 7½ins. high. £1,150

Rare Jacobite wine glass engraved with badge of Society of Sea Sergeants. £1,850

Flute glass with the Arms of Charles II and James II engraved in diamond point, made of soda glass and almost certainly Dutch, 15½ins. high. £2,300

18th century English wine glass etched in diamond point with Jacobite verses, 15.9cm. high. £2,700

An heraldic goblet decorated by Beilby of Newcastle, 8½ins. high. £3,100

Beilby armorial goblet inscribed 'W. Beilby Jr.', dated 1762, 8¾ins. high. £19,500

INDEX